DOT/FAA/AM-10/16
Office of Aerospace Medicine
Washington, DC 20591

Developing Proactive Methods for General Aviation Data Collection

Scott Shappell,[1] Carla Hackworth,[2]
Kali Holcomb,[2] John Lanicci,[3]
Massoud Bazargan,[3] Jaclyn Baron,[1]
Rebecca Iden,[1] Daniel Halperin[3]

[1]Clemson University
 Clemson, SC 29634

[2]Civil Aerospace Medical Institute
 Federal Aviation Administration
 Oklahoma City, OK 73125

[3]Embry-Riddle Aeronautical University
 Daytona Beach, FL 32114

November 2010

Final Report

NOTICE

This document is disseminated under the sponsorship
of the U.S. Department of Transportation in the interest
of information exchange. The United States Government
assumes no liability for the contents thereof.

Technical Report Documentation Page

1. Report No. DOT/FAA/AM-10/16	2. Government Accession No.	3. Recipient's Catalog No.
4. Title and Subtitle Developing Proactive Methods for General Aviation Data Collection		5. Report Date November 2010
		6. Performing Organization Code
7. Author(s) Shappell S,[1] Hackworth C,[2] Holcomb K,[2] Lanicci J,[3] Bazargan M,[3] Baron J,[1] Iden R,[1] Halperin D[3]		8. Performing Organization Report No.
9. Performing Organization Name and Address [1]Clemson University; Clemson, SC 29634 [2]FAA Civil Aerospace Medical Institute; Oklahoma City, OK 73125 [3]Embry-Riddle Aeronautical University; Daytona Beach, FL 32114		10. Work Unit No. (TRAIS)
		11. Contract or Grant No.
12. Sponsoring Agency name and Address Office of Aerospace Medicine Federal Aviation Administration 800 Independence Ave., S.W. Washington, DC 20591		13. Type of Report and Period Covered
		14. Sponsoring Agency Code
15. Supplemental Notes Work was accomplished under approved task AM-HRR 521.		
16. Abstract **Introduction.** Over the last 20 years, nearly 40,000 general aviation (GA) aircraft were involved in accidents, roughly 20% of which were fatal. To address this safety concern, scientists have often relied on accident data. Because of the rare nature of accidents, commercial aviation incident and near miss data may prove to be useful sources of safety information. In one such study, the National Transportation Safety Board interviewed GA pilots that were flying near a weather-related accident in pursuit of a different perspective than that of the accident pilot. Interviewing GA pilots about their own weather-related event may provide similar benefits. **Method.** To understand factors leading GA pilots to encounter adverse weather conditions, pilots involved in an adverse weather encounter were interviewed using a one-hour structured interview. The interview was developed using surveys utilized by National Aeronautics and Space Administration and the Federal Aviation Administration (FAA). In total, 27 pilots who experienced an adverse weather encounter were interviewed, of which 25 were included in the final analysis. **Results.** Previous studies conducted by the FAA and others found many GA accidents involving flight into adverse weather were categorized as a willful disregard for the rules and regulations of safety; violations as defined by the Human Factors Analysis and Classification System. Contrary to what the accident record seems to suggest, flight into adverse weather may also be influenced by the lack of appreciation/understanding of the hazards associated with adverse weather. Perhaps some encounters with adverse weather were motivated by outside influences or exacerbated by some manner of mechanical failure that may have led to the willful acceptance of unnecessary hazards. **Conclusions.** These data suggest that current beliefs surrounding flight into adverse weather by GA pilots may be incomplete. The data presented here suggest that additional effort should be placed in training, both *ab initio* and recurrent. Emphasis should be placed on ensuring a full understanding of the adverse impact of weather, including the recognition of instrument meteorological conditions, icing, convective events, etc. Likewise, with the proliferation of commercial weather products and on-board weather equipment, it may be time to move toward some form of standard weather package that all pilots would review before flying.		
17. Key Words General Aviation, Aviation Weather Encounters, Weather Knowledge, Pilot Education and Training		18. Distribution Statement Document is available to the public through the Defense Technical Information Center, Ft. Belvoir, VA 22060; and the National Technical Information Service, Springfield, VA 22161
19. Security Classif. (of this report) Unclassified	20. Security Classif. (of this page) Unclassified	21. No. of Pages 23	22. Price

Form DOT F 1700.7 (8-72) Reproduction of completed page authorized

ACKNOWLEDGMENTS

We are indebted to the many Flight Standards District Office employees that provided assistance beyond their normal duties to identify pilots as requested. As mentioned, this often involved additional searching on their part; we are very appreciative of their efforts and grateful for their commitment to aviation safety.

Research reported in this paper was conducted under the Flight Deck Program Directive / Level of Effort Agreement between the Human Factors Research and Engineering Group (AJP-61), Flight Standards Service – General Aviation and Commercial Division (AFS-810), and the Aerospace Human Factors Division (AAM-500) of the Civil Aerospace Medical Institute. This study was supported by FAA Cooperative Agreement Number 06-G-002.

CONTENTS

INTRODUCTION 1
METHOD 2
    Participants 2
    Instrument 3
    Aircraft Demographics 3
    Pilot Demographics 3
    Event Information 3
    Preflight Planning 3
    Enroute decision-making 4
    Procedures 4
    Weather Information 4
RESULTS 4
    Pilot information 4
    Pilot Certification 4
    Flight Experience 4
    Physical and Mental Health 5
    Aircraft information 5
    Preflight planning 5
    Route Planning 5
    Weather Planning 6
DISCUSSION 9
CONCLUSIONS AND LESSONS LEARNED 12
REFERENCES 14
APPENDIX A: Pilot Questionnaire A1

DEVELOPING PROACTIVE METHODS FOR GENERAL AVIATION DATA COLLECTION

INTRODUCTION

Operating a general aviation (GA) aircraft, particularly when flying cross-country, requires dedicated planning and decision-making involving such issues as the route of flight, terrain and obstacles one might encounter, and the threat of adverse weather. Some of these challenges remain fixed (e.g., terrain and other obstacles) and are relatively straightforward to address, while others, such as weather, are dynamic and can pose unique challenges to pilots. Indeed, even the best flight plans can prove inadequate or incomplete where the volatile nature of adverse weather is concerned.

Even with its unpredictable nature, pilots are still expected to fully understand the weather enroute and to avoid severe weather when planning and executing their flight (National Transportation Safety Board, 2006). Specifically, Title 14 Code of Federal Regulations (CFR) Part 91.103 mandates that when conducting a flight under instrument flight rules (IFR) or away from the airport pilots should become familiar with weather reports and forecasts (Federal Aviation Regulations/Aeronautical Information Manual, 2009).

Despite the regulations that require pilots to fully understand and appreciate the risk associated with adverse weather, weather-related GA accidents continue to occur. To put this in perspective, over the last 20 years, nearly 40,000 GA aircraft have been involved in accidents; of these, roughly 20% involved fatalities (Shappell & Wiegmann, 2009). Notably, many of those fatal accidents involved encounters with adverse weather (Detwiler, Boquet, Holcomb, Hackworth, Wiegmann, & Shappell, 2006; Wiegmann, Boquet, Detwiler, Holcomb, Faaborg, & Shappell, 2005; NTSB, 2005).

So, why would a pilot fly into adverse weather, particularly when the consequences can be so dire? Unfortunately, the answer continues to be debated among academics and regulators, alike. Some of the more prevailing explanations have recently been summarized by Wiegmann, Talleur, and Johnson (2008) and include:

- Lack of knowledge and experience (Goh & Wiegmann, 2002; O'Hare & Owen, 2002; Wiegmann, Goh, & O'Hare, 2002)
- Poor pre-flight planning (Wiggins, Stevens, Howard, Henley, & O'Hare, 2002; Wiggins & O'Hare, 2003)
- Limited weather-evaluation skills (Goh & Wiegmann, 2001; Wiegmann, et al., 2002; Burian, Orasanu, & Hitt, 2000; Wiggins & O'Hare, 2003)
- Low risk perception/overconfidence (Goh & Wiegmann, 2001)
- Poor in-flight planning (Detwiler et al., 2006; Stokes, Belger, & Zhang, 1990; Knecht, 2005)
- Inability to execute a 180 degree turn (Wiegmann & Goh, 2003; AOPA, 2004)

Some have even suggested that encounters with adverse weather represent the willful disregard for the rules and regulations (Wiegmann et al., 2005). After all, it is incomprehensible to some that a pilot who sees threatening weather in his path would elect to continue into it if a safer alternative is available. The problem with this latter view is that it is based upon accident data that are arguably limited, particularly when the pilot is fatally injured, and the exact causes of accidents are based as much on conjecture and expert opinion as they are on fact.

Despite the uncertain nature of fatal accident data, aviation safety has historically been driven by accident databases archived and maintained by organizations like the National Transportation Board (NTSB) and Federal Aviation Administration (FAA). However, at some point, relying solely on accident data will yield fewer and fewer positive interventions – particularly as the historical accident rates are driven down to the level of rare events.

This is precisely what has happened within the commercial aviation industry. As a result, emphasis within commercial aviation has moved from accident data to more proactive and normative data sources and risk analysis. For instance, programs like flight operations quality assurance (FOQA), the aviation safety action program (ASAP), and line operations safety audits (LOSA) have been developed and embraced by the commercial aviation industry.

At a minimum, these proactive and normative data sources have expanded our understanding of the human factors associated with commercial aviation accidents and incidents and have broadened how accident/incident causation is viewed. However, even established data-collection tools like these continue to be modified and have yet to be leveraged throughout the aviation industry – particularly GA where equipment, programs, and access are less regulated and standardized.

Left with few alternatives, many studies of GA flight into adverse weather have utilized simulation, surveys, and structured interviews. One particularly unique study of interviewed pilots was that they were flying in the vicinity of, but not directly involved in, a weather-related accident (NTSB, 2005). This study was conducted to render a different perspective than that of the accident pilot, who in most cases is fatally injured. This represented a significant departure from traditional accident studies in that the pilots interviewed were not directly involved in the accident but were exposed to similar environmental conditions (e.g., weather and terrain). Among the more notable findings were:

- A pilot who starts flying earlier in life is at lower risk of being involved in a weather-related GA accident than those who start flying when they are older.
- Chronological age at the time a pilot obtains a license to fly is a better predictor of future accident involvement than age at time of flight.
- Periodic training and evaluation may be necessary to ensure that pilots maintain weather-related knowledge and skills.
- GA pilots routinely consult other sources of aviation weather to obtain information that is not currently available from a standard weather briefing (i.e., Flight Service Station).

As useful as the NTSB study was, most would agree that information from a pilot involved with the adverse weather encounter would be even more valuable. The challenge is to identify a source of data involving actual pilot encounters with adverse weather.

One relatively untapped source of information is near-miss data. By definition, these are not accidents and, as such, the pilots involved are alive and can presumably recount their decision process. Of course, this assumes that those issues that influence performance during accidents also influence near misses in a similar way.

One such "near-miss" database already exists. The Aviation Safety Reporting System (ASRS) is an anonymous database that relies on self-reported incidents and near-miss data. However, ASRS reports are not necessarily focused on any specific type of incident or near-miss. Moreover, they require the individual to self-report, which may constrain submitted information, and certainly does not go to the level of detail required for a full understanding of pilot decision-making (Macrae, 2009). In sum, what is needed is a better understanding of exactly how and why a GA pilot would encounter adverse weather straight from the source – the pilot who made the decision and experienced the weather encounter.

METHOD

Although there are more traditional ways to identify pilots who encountered adverse weather and survived (e.g., NTSB accident records and ASRS reports), this study employed two previously untapped resources: 1) the Administrators' Daily Alert Bulletins that identify incidents, accidents, and pilot deviations on a daily basis, and 2) air traffic control flight assists. The latter occur when pilots request the assistance of air traffic controllers to navigate their aircraft or when an air traffic controller initiates an interaction with a pilot for safety purposes. Many of these interactions involving flight through or around adverse weather are reported to responsible FAA officials and were of particular interest. Monitoring these events is consistent with FAA order 8000.1, Safety Management System Doctrine, which states:

> The safety assurance process continually assesses activity to identify new hazards and to ensure risk controls achieve their intended objectives throughout the system life cycle. New hazards may be those not identified during the SRM process or those introduced by the risk controls. This includes assessment of the need for new risk controls or to eliminate or modify risk controls that are ineffective or are no longer needed due to changes in the operational environment. Every SMS includes a process for continuously monitoring systems of interest to identify new hazards or the need to change risk controls or other risk management responses. These monitoring activities apply to an SMS whether the operations are accomplished internally or outsourced. (FAA Order 8000.1, Section 2-5.3.a.).

With both data sources, attention was paid to events involving visual flight rules (VFR) into instrument meteorological conditions (IMC), convective and icing encounters, and instrument-rated pilot control difficulties in actual IMC.

Participants

More than 175 events were reviewed, of which 115 were included as candidates for this study. Scientists from the FAA Civil Aerospace Medical Institute (CAMI) contacted the Flight Standards District Office (FSDO) aviation safety inspectors (ASIs) responsible for the candidate events to describe the project objectives and request permission to communicate with the involved pilot. Note that in some cases no investigation had been conducted prior to CAMI contact. In total, more than 160 ASIs from roughly 55 different FSDOs were contacted.

From the ASI inquiries, approximately 50 potential participants were identified and contacted via the telephone, E-mail, or postal mail, depending on the type of contact information received. After being briefed on the purpose of the project, including the anonymous nature of their responses, pilots who agreed to participate were provided a consent form and scheduled for a telephone interview.

Twenty-seven pilots who experienced an adverse weather encounter were subsequently interviewed. Two pilot interviews were eliminated because one was interviewed during initial beta testing, and the other did not meet our research needs. This left 25 pilot interviews in the final analysis.

Instrument

The roughly one-hour interview was developed using surveys previously employed by the National Aeronautics and Space Administration (NASA) and the FAA (NASA, 2007; Knecht, 2008a and 2008b). The structured interview included several items on basic aircraft and pilot demographics, a description of the event, preflight planning, and weather decision-making. A brief description of each section of the structured interview follows:

Aircraft Demographics

Pilots were asked standard demographic questions, such as the type of aircraft they were flying during the weather encounter and whether they leased, partially, or fully owned the aircraft.

Pilot Demographics

In addition to traditional demographic questions like education, profession, gender, and age, several items regarding piloting experience and training were asked of the pilots. Specifically, they were asked:
- When they first learned to fly
- The date of their most recent flight instruction
- The date of their most recent biennial flight review
- Class of medical certificate they held and whether any waivers or limitations were attached
- Ratings and certificates they held (e.g., single engine land, certified flight instructor, and instrument rating)
- Flight hours flown before the adverse weather encounter (total and 90 days prior to the event)
- Number of cross-country flight hours flown before the adverse weather encounter (total and 90 days prior to the event)
- Instrument flight hours flown before the adverse weather encounter (simulated and actual instrument flight)
- Type of equipment (e.g., weather-avoidance, terrain-avoidance, autopilot, GPS, and de-icing) onboard the aircraft; if onboard, was it used during the flight and had the pilot received formal training for its use
- Whether they participated in FAA-sponsored WINGS programs
- Whether they obtained a good night's sleep before the flight
- Whether they were suffering from any illness the day of the flight

Event Information

Pilots were asked to describe their weather encounter in detail. Several additional demographic questions related to the flight were also asked, including:
- Purpose of the flight (i.e., business or pleasure)
- Had they flown the route before and if so, how many times?
- Planned length of flight
- Departure airport
- Approximate temperature at departure
- Time of departure
- Planned destination airport
- Type of approach planned for destination (i.e., Visual, Cat I, II, or III)
- Planned alternate destination airport
- Whether they communicated with ATC or any other services (e.g., Flight Service Stations) while enroute
- In-flight services requested (e.g., an emergency climb/descent, vectors to an airport or visual meteorological conditions (VMC), pilot reports (PIREPS), weather updates, or instrument approach procedures)
- Lighting conditions (e.g., daylight, night, dusk, dawn)
- Weather conditions (i.e., IMC, VMC, marginal VMC) at departure, enroute, and at the destination
- Time of arrival at their intended or alternative destination

Preflight Planning

Of particular interest in this study was the method of preflight weather planning employed by the pilots. Pilots were asked to describe their normal method of preflight planning and whether it was different the day of the weather encounter. They were also asked:
- Whether they attempted to obtain pre-flight weather information
- The last time they checked weather before flight
- Which weather providers they typically use when planning a flight (e.g., National Weather Service, flight service stations, direct user access terminal, and hazardous in-flight weather advisory service) and if those were used before the incident flight
- Were any attempts to obtain pre-flight weather information unsuccessful and if so, why?
- Whether the actual weather experienced was the same as, better, or worse than predicted when they departed, enroute, and at their planned destination
- Whether they considered the geography along the flight route (e.g., mountains and rising terrain)

Enroute decision-making

Because all participants had encountered adverse weather, several questions were asked regarding their enroute decision-making. Specifically, pilots were asked:

- Whether they were under any pressure to get to their destination
- Whether they were aware of their course and location throughout the flight
- Whether there were any distractions in the cockpit
- Whether all the equipment was working during the flight
- What was their level of weather awareness during the flight
- What was their thought process upon entering the weather
- How did they feel upon encountering the weather
- Had they ever experienced similar weather conditions
- Whether they considered returning to their departure point or performing a 180 degrees to avoid weather
- Whether they had to divert from their planned destination and if so, was it before or after encountering the weather
- Whether they encountered the weather after reaching the half-way point of the flight
- What they did to prevent the encounter with weather from being worse
- Whether they would have done anything differently

Procedures

The structured interview was conducted using a conference call involving scientists from CAMI, Clemson University, Embry-Riddle Aeronautical University (ERAU), and the pilot participant. After reviewing the purpose of the interview and confirming the anonymous nature of the study and other elements of the informed consent, the roughly one-hour structured interview was conducted (see Appendix A for the actual structured interview questions). Responses were independently recorded for later comparison. Note that the interview was not audio-taped, only hand-written or typed notes of the interview were produced.

After the interview was completed, recorded responses were compared to ensure the accuracy of the record. A single consensus file was then sent to the pilot for review. This gave the pilot-participant an opportunity to ensure the accuracy of the recorded responses and to modify or clarify any responses given during the interview. Once the interview had been reviewed and accuracy ensured by all parties, the responses were combined with the others contained within a de-identified master database of pilot interviews.

Weather Information

In addition to pilot interview data, information was obtained regarding the atmospheric conditions during the time and at the location of each flight and weather encounter. Archived meteorological data associated with each event in the final database were obtained from the National Climatic Data Center. Relevant aviation weather reports were reviewed including:

- Routine aviation meteorological reports (METARs)
- Terminal aerodrome forecasts (TAFs)
- Airman's meteorological information (AIRMETs)
- Significant meteorological information/ advisories (SIGMETs)
- Precipitation reflectivity fields from the National Weather Service's Doppler Radar Network

The METARs and TAFs were collected for the departure, destination, and encounter/diversion times and locations in each case. The AIRMETs, SIGMETs, and radar data were collected for the routes/times of flight, and the encounter times and locations were also documented.

RESULTS

Pilot information

All pilots interviewed were male, ranging in age from 20 to 72 years old, with an average age of 50 years. Ten pilots (40.0%) were between the ages of 40 and 59, and nine (36.0%) were 60 and over. All pilots had completed some college coursework and almost half (48.0%) had received graduate level training. Many of the pilot participants were employed as professional pilots or as business owners.

Pilot Certification

All pilots had a basic pilot certification of airplane single engine land and were medically certified to fly. Over half (60.0%) held a Class III (private pilot only) medical certificate, with the remaining pilots holding either a Class II (commercial, non-airline duties, and private pilot; 24%) or Class I (scheduled airline) medical certificate (16%). In addition to the single engine land rating, 36.0% of pilots held a multi-engine rating, 24.0% held a commercial rating, and 20.0% were certified flight instructors. Additionally, most of the pilots (76.0%) were instrument rated.

Flight Experience

Although flight experience varied greatly among the pilots, our pilot cohort can be considered experienced pilots. A summary of their flight experience is presented in

Table 1. Pilot participant flight experience

	Median	Minimum	Maximum
Total flight hours	1,100.0	130.0	20,000.0
Total hours in event aircraft make/model	300.0	4.0	5,000.0
Total hours in last 90 days	40.0	2.0	250.0
Cross-country hours	700.0	35.0	14,000.0
Cross-country hours in last 90 days	36.0	0.0	250.0
Actual instrument hours (n=16)	164.0	4.0	2,970.0
Simulated instrument hours (n=13)	45.0	10.0	242.0
Total instrument hours (n=15)	145.0	7.0	3,000.0

Table 1. As can be seen, experience varied greatly, ranging from 130 total hours to 20,000 total hours. The median for the group was 1,100 total hours of experience. Note that because of the wide range in flight hours, the median was viewed as the best measure of central tendency. Actual instrument flight time also varied markedly, ranging from 4 to 2,970 hours with a median of 164 hours.

Most of the pilots (56.0%) first learned to fly within 10 years of the adverse weather encounter. More than two-thirds of the pilots (68.0%) received some degree of formal flight instruction within the year prior to the weather encounter. However, two (8.0%) of the pilots had not received formal flight instruction in more than a decade.

The majority of the pilots (80.0%) had completed at least one biennial flight review (BFR), and most (70%) had completed a BFR within 1 year of the weather encounter. It was also interesting that almost half (48.0%) of the pilots had participated in FAA-sponsored WINGS programs, which provide voluntary training to enhance pilot proficiency and safety.

Physical and Mental Health

The majority of the pilots (92.0%) reported feeling well on the day of the weather encounter. Those who were feeling worse than normal that day reported allergies or were recovering from being sick a few days prior to the flight. However, all had obtained a good night's sleep the night before the flight. Most did not feel pressured to get to their destination; however 20.0% of the pilots reported having some form of motivation to arrive at their planned destination (i.e., wanting to get home as soon as possible.)

Aircraft information

Not surprisingly, a variety of aircraft was represented in our sample. Most of the aircraft were owned by the pilot (60.0%). The remaining pilots flew either a company-owned (24.0%) or rental (16.0%) aircraft.

Several types of aircraft were represented in this study. While most were single-engine, propeller-driven aircraft (e.g., Cessna, Beechcraft; 84.0%), there were a few multi-engine aircraft (e.g., KingAir and Piper; 12.0%) and one Learjet. The top-three aircraft manufacturers were Cessna (32.0%), Piper (20.0%), and Mooney (16.0%).

Almost all (96.0%) of the pilots had at least one GPS inside the aircraft. Approximately two-thirds (62.5%) of the aircraft had terrain avoidance equipment and over half (56.0%) had on-board weather-avoidance equipment. Almost half (44.0%) of the aircraft had de-icing equipment, and most (79.2%) were equipped with an autopilot.

Preflight planning

Preflight planning is critical in every flight. Included in a comprehensive preflight are such things as the aeronautical integrity of the aircraft, engine maintenance, fuel quantity, route planning, alternate airports, and weather. While all are important, route planning, alternate airports, and preflight weather planning were of greatest concern in this study.

Route Planning

Our sample included routes across a variety of geographical locations; however, most occurred in the Southern and Northwest Mountain regions, as defined by the FAA. A regional map of where the weather encounters occurred is presented in Figure 1.

The majority of the flights (75.0%) took place during the day and were conducted mainly for business (40.0%) and pleasure (48.0%); the remaining flights were for multiple purposes (8.0%) and training (4.0%.) Most of the flights (80.0%) were planned to last up to 5 hours, with an average flight time of just more than 2 hours. The majority (76.0%) of pilots had planned a visual approach to landing at their destination.

All pilots considered the geography that they would encounter enroute and approximately two-thirds (64.0%) had flown the route before. Likewise, most (64.0%) of the pilots had planned an alternate airport in advance of

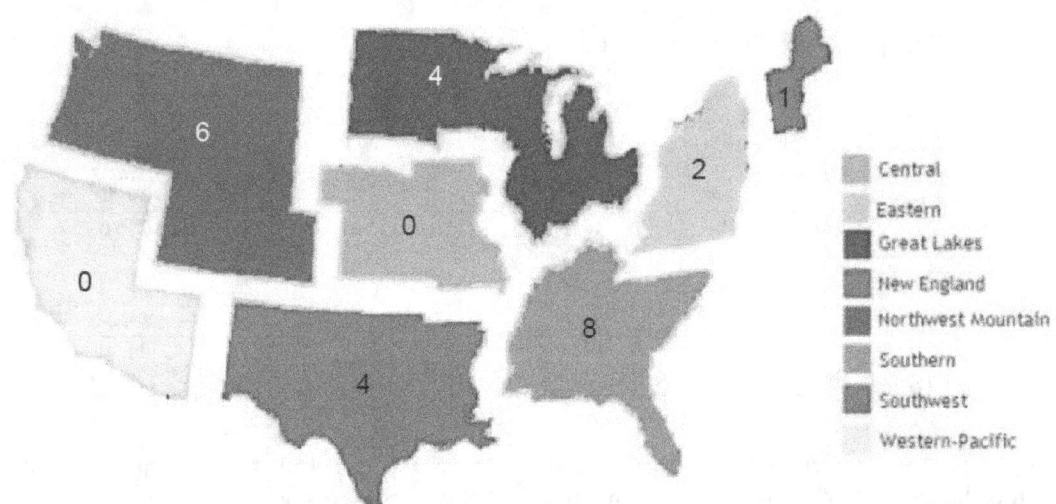

Western Pacific	Northwest Mountain	Southwest	Central	Great Lakes	Southern	Eastern	New England
Arizona California Hawaii Nevada	Alaska Colorado Idaho Montana Oregon Utah Washington Wyoming	Arkansas Louisiana Oklahoma New Mexico Texas	Iowa Kansas Missouri Nebraska	Illinois Indiana Michigan Minnesota North Dakota Ohio South Dakota Wisconsin	Alabama Florida Georgia Kentucky North Carolina South Carolina Tennessee Mississippi	Delaware Maryland New Jersey New York Pennsylvania Virginia West Virginia District of Columbia	Connecticut Maine Massachusetts New Hampshire Vermont
0	6	4	0	4	8	2	1

Figure 1. Location of weather encounters using FAA defined regions.

Table 2. Number of pilots using the specified weather services on the day of the weather encounter and in general (n=22)[1]

	Number of Pilots	
Number of sources	Used day of encounter	Typically used
1	0	1
2	4	0
3	6	4
4	4	3
5	2	2
6	2	3
7	4	3
8	0	4
9	0	2

[1]Note that the sample size is 22 because this item was added after the first interviews were complete.

takeoff in the event of encountering difficulties enroute. While alternate airports may not be important when flying locally, most of these flights involved cross-country flying, and alternate airports should normally be considered.

Weather Planning

All pilots had obtained some type of weather information prior to departure on the day of the weather encounter, and most (76.0%) accessed weather information less than 30 minutes before departure. The sources that the pilots used to access weather information varied. However, only three pilots experienced difficulty accessing weather information in their typical manner; in those cases, weather information was obtained from alternate sources.

Pilots were also asked about the weather services they used during the day of the encounter and which services they typically used. However, after a few interviews, the questions were modified to increase the specificity of the pilot responses. Table 2 presents the number of weather services that the pilot participants used, both typically and on the day of the weather encounter.

The data revealed that the pilots in this study "typically" used between one and nine weather sources to access weather information, with an average of 5.6 sources.

Table 3. Weather services used that day and typically used, by number of pilots (n=22)[2]

	Used Day of Encounter	Typically Used
National Oceanic & Atmospheric Administration	14 63.6%	16 72.7%
Flight Service Station	15 68.2%	17 77.3%
DUATS	11 50.0%	12 54.5%
Commercial vendor	11 50.0%	17 77.3%
Hazardous In-flight Weather Advisory Service	4 18.2%	9 40.9%
Transcribed Weather Broadcast	3 13.6%	5 22.7%
Pilots Automatic Telephone Weather Answering Service	0 0.0%	1 4.5%
En Route Flight Advisory Service	6 27.3%	10 45.5%
Weather Channel	12 54.5%	16 72.7%
Other Pilots	6 27.3%	13 59.1%

[2]Note that the sample size is 22 because this item was added after the first interviews were complete.

Table 4. Weather forecast

	VMC	Marginal VMC	IMC
Departure Airport	16 (72.7%)	3 (13.6%)	3 (13.6%)
En Route	8 (36.4%)	6 (27.3%)	8 (36.4%)
Destination Airport	13 (61.9%)	5 (23.8%)	3 (14.3%)

Note n= 22 data for the departure and en route forecasts and n=21 for the destination forecast

Indeed, roughly half (54.5%) of these pilots typically used six or more weather sources.

That being said, on the day of the weather encounter, the pilots "actually" used slightly fewer weather sources– between two and seven weather sources to obtain weather information, with an average of 4.2 sources. Moreover, on the day of the weather encounter, only 27.3% of the pilots used six or more weather sources – roughly half of what was reported as "typically" used.

The type(s) of weather service(s) used by pilots typically and on the day of the weather encounter were very similar (Table 3). For instance, the top three weather providers used on the day of the weather encounter were Flight Service Stations, National Oceanic & Atmospheric Administration (NOAA), and the Weather Channel. However, that is not the entire story, as all of the weather providers, except for one category, were used less frequently on the day of the event, compared to typical usage rates.

As can be seen from Tables 4 and 5, most flights were forecast to depart in VMC with the actual weather at departure the same or better than predicted. However, most of the pilots (60.0%) encountered adverse weather enroute, with almost two-thirds of the encounters (64.0%) occurring past the halfway point of the flight.

Instrument meteorological conditions (IMC) and icing were the two most frequently encountered weather conditions. Table 6 shows all types of weather encountered.

Most of the pilots (96.0%) reported being aware of the adverse weather and where it was headed. However, the majority of the pilots (96.0%) ended up entering the weather anyway. In fact, only a little more than half of those pilots (58.3%) even considered performing a 180 degree turn to return to the departure airport or an alternate airport.

As expected, the majority of pilots (92.0%) deviated at some point from their planned route or altitude. The two pilots who did not deviate from their planned routes were

Table 5. Actual weather experienced as compared with forecasted weather[3]

	Better	Same	Worse
Departure Airport	3 (14.3%)	16 (76.2%)	2 (9.5%)
En Route	2 (9.5%)	6 (28.6%)	13 (61.9%)
Destination Airport	0 (0%)	4 (44.4%)	5 (55.6%)

[3] *Four pilot data points were missing for departure and en route (N=21). Many pilots did not make it to the intended destination airport (N=9).*

Table 6. Type of weather encountered[4]

	Percentage
Thunderstorm	17.0%
Icing	41.7%
IMC	50.0%
Non-Convective Turbulence	4.2%
MVMC	8.0%

[4] *Pilots encountered multiple types of weather; therefore, the sum of percentages equal more than 100%.*

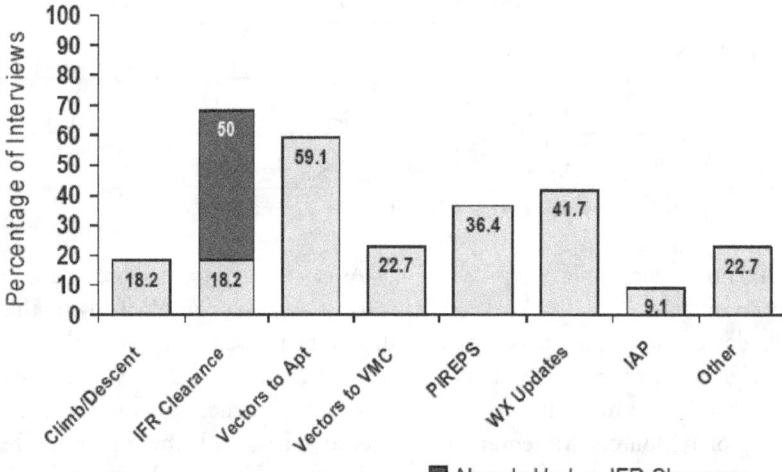

Figure 2. In-flight services requested.

very close to their intended destination or on the taxiway at the destination airport when the weather encounter occurred. However, the majority who deviated (73.9%) did so <u>after</u> encountering the weather. Even though a majority of the participants deviated, 87.5% were aware of where they were relative to their course and location.

The majority of the pilots (76.0%) reported that they had previously flown in similar conditions.

Because many of the pilots were identified using air traffic controller flight assists, it is not surprising that all but one pilot (96.0%) communicated with air traffic control. A variety of in-flight services were requested of air traffic control and flight service. While a few (12.0%) pilots requested as many as five services, the majority of the pilots made only one (28.0%) or two (24.0%) requests in-flight. As illustrated in Figure 2, the most commonly requested in-flight services were vectors to an airport, weather updates, and PIREPS. Interestingly, half of the pilots already had IFR clearance for the flight, and around one-third (36.4%) of the remaining pilots requested IFR in-flight.

It is interesting to note that more than half of these weather encounters (56.0%) were exacerbated by equipment failures, which can have dramatic impacts on pilot decision-making. For example, in a series of studies involving GA pilots, Beringer & Harris (2007) were able to show that automation-related (i.e., runaway pitch-trim up and down, roll servo failure, roll sensor failure, and pitch drift up) and mechanical failures (e.g., failed attitude indicator) can result in the a variety of piloting errors resulting in the loss of the aircraft. Put simply, the presence of a mechanical malfunction increases the likelihood that an error will occur.

DISCUSSION

Human decision-making in general, and pilot decision-making in particular, is largely dependent upon three key elements: 1) information – is the information accurate and timely; 2) knowledge – does the individual have the requisite knowledge to utilize the information present; and 3) experience – has the individual obtained sufficient experience with a given situation or similar situations to make a correct assessment (Patterson & Shappell, in review). The likelihood that a decision will be successful is markedly reduced if any of these three components are absent or lacking. It can also be argued that motivation influences to what extent the three elements (information, knowledge, and experience) will be considered in the decision-making process.

With this in mind, the 25 weather encounters were classified into one of five categories using the interview data and narrative summary of the encounter provided by the pilot participants. The categories utilized were:

- Motivation. Referred to by some as "get-home-itis," some pilots feel pressured to get to their destination. That pressure can be either external or internal.
- Lack of complete weather information. In some instances, pilots do not receive complete weather information and/or do not recognize that the information provided is insufficient.
- Conflicting weather information. In some instances, sources of weather information may be in variance. For instance, on-board NEXRAD radar may be indicating a break in convective activity, while ATC and/or Flight Service may report the same area as IMC.
- Lack of appreciation/understanding of the weather. Some pilots may not understand the implications of the weather, or if they do, they may not appreciate the threat to flight safety that adverse weather poses. In effect, these pilots lacked a practical strategy for managing the weather hazard they faced.
- Not/applicable. This category was included to capture those encounters that could not be classified within the previous four categories (e.g., encounters due to mechanical failures).

As can be seen in Figure 3, roughly one-fourth of the adverse weather encounters were driven by the motivation of pilots to get to their destination. These instances may explain at least some of the violations that have been identified in the accident record by Wiegmann et al. (2005). However, the majority of adverse weather

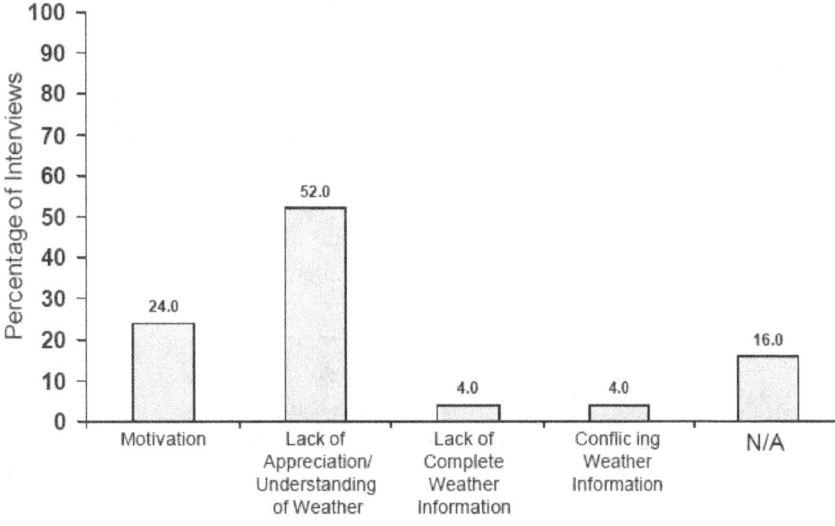

Figure 3. Explanations for why pilots fly into adverse weather.

Table 7. Decision category by type of weather

Hazard	Lack of Appreciation/ Understanding of Weather	Motivation	Conflicting Weather Information	Lack of Complete Weather Information
IMC	5	4	0	1
Icing	6	1	1	0
Non-convective Turbulence	0	0	0	1
Convective Weather	4	0	0	0
Marginal VFR	1	1	0	0

encounters were due to a lack of understanding/appreciation of the weather conditions (52.0%). Surprisingly, few adverse weather encounters were caused by incomplete or conflicting weather information (8.0%). Note that four weather encounters did not fit into any of the aforementioned categories. After revisiting the uncategorized (N/A) encounters, all but one could be attributed to a technological malfunction (e.g., a malfunctioning onboard weather radar).

What's more, the lack of understanding/appreciation of the weather did not seem to be isolated to any specific type of adverse weather event (Table 7). That is, the lack of appreciation/understanding of weather was spread across weather hazards. In contrast, nearly all adverse weather encounters that involved misplaced motivation were coupled with general IMC, rather than the more dangerous convective weather. This suggests that when faced with exceptionally bad weather, motivation may not have as much influence on pilot behavior as originally thought.

In general then, it would appear from this study that rather than willfully disregarding the rules, many of the pilots simply committed decision errors, as described within HFACS[1] (Wiegmann & Shappell, 2003). Indeed, these decision errors were consistent with many of the explanations of why pilots fly into adverse weather, provided by Wiegmann et al. (2008). Explanations such as the lack of knowledge and experience, poor pre-flight planning, limited weather evaluation skills, and poor in-flight planning all would contribute to decision errors and seem to corroborate the findings presented here.

Does this mean that previous findings (e.g., Wiegmann et al., 2005; Detwiler et al., 2006) were flawed that suggested that many GA weather-related accidents are due to violations? Not necessarily. It could be that there are fundamental differences between fatal and non-fatal encounters with weather. Indeed, data from those studies suggest that most fatal weather encounters are associated with violations, while non-fatal weather encounters are more indicative of decision errors (Wiegmann et al., 2005; Detwiler et al., 2006). Obviously, the majority of events presented here did not involve accidents on fatalities.

It can also be argued that the experience level of the pilots in this study was considerably higher than those GA pilots examined in previous analysis of the NTSB accident database (Goh & Wiegmann, 2002). Roughly 40% of the pilots in the analysis of the NTSB accident database possessed an instrument rating, while a much larger proportion (76%) of the pilots in this study were instrument rated.

On the other hand, and perhaps more troubling, it could be that the lack of available information following fatal weather encounters may bias investigators to resort to explanations more indicative of the willful disregard for the rules and regulations (i.e., violations). Without substantiating information from occupants of the aircraft or those with intimate knowledge of the pilot, field investigators are often left with little more than superficial descriptions of accidents that may lack the specificity required to rule out violations. Unfortunately, it will never truly be known what decisions were made in the cockpit when the pilots suffer fatal injuries. Indeed, there may be more to the story. This study showed how pilot decision-making and the lack of understanding of key weather information could easily lead to fatal accidents. For many, it was a matter of good fortune that under different circumstances (i.e., different amount of fuel, time of day, lack of ATC assistance, no hole in the cloud layer), the outcome could have resulted in a serious or fatal accident.

It is noteworthy that the findings presented here seem to support those reported by the NTSB (2005). Recall that NTSB researchers interviewed pilots who were in the

---

[1] The Human Factors Analysis and Classification System (HFACS) is a theoretically based tool for investigating and analyzing human error associated with accidents and incidents. HFACS was developed by expanding upon Reason's (1990) "Swiss-cheese" model of human error. The basic premise underlying this taxonomy is that an accident is generally a series of events. The sequence of preceding events are purported to include both latent and active errors. The HFACS taxonomy defines 19 causal categories of human error whereby the accident's progression is categorized. These categories are housed within four levels (i.e., unsafe acts, preconditions for unsafe acts, unsafe supervision, and organizational influences) of human error. See Wiegmann and Shappell (2003) for an in-depth description.

general vicinity of those whose encounter with adverse weather resulted in a fatal accident. In contrast, the pilots interviewed in this study *actually encountered* the weather.

Consistent with the "accident pilots" in the NTSB study, many of the pilots interviewed in this study were older and started flight training later in life – both areas of concern identified by the NTSB (NTSB, 2005). Likewise, the data presented here regarding use of weather resources by pilots are consistent with that observed for pilots flying in the vicinity of the accident pilots. Specifically, the data presented here suggest that pilots utilize a variety of weather sources and that no identifiable "standard" approach to weather pre-flight planning appears to exist within GA. Unfortunately, it remains largely unknown what specific weather sources the accident pilots in the NTSB study utilized.

Given the proliferation of commercial weather products and on-board weather equipment, it may be time to move toward some form of standard weather package that all pilots would review before flying. Exactly what products/equipment should be used remains to be defined; however, it is clear that some weather products and providers are considered more useful and/or are accessed more often than others (NASA, 2007). For instance, FSS, DUATS, and NOAA/NWS, were all utilized in both the NASA study and here. Similarly, Knecht (2008a) surveyed 230 GA pilots from locations across the country (CA, OK, ND, IL, FL) and found that the FAA standard briefing was used on a majority of flights, followed by the NWS/NOAA and DUATS. Beringer and Schvaneveldt (2002) found close agreement, for the most part, between the ratings of weather information items by both novice and experienced pilots. They recommended, based upon these findings, that one could develop algorithms, not unlike Pilot's Associate, that would prioritize weather information by phase of flight for presentation on flight-deck displays, thereby potentially reducing display clutter and emphasizing salience.

Of course, simply accessing weather data does not necessarily mean that one paid attention to relevant information or that the information was even fully understood. It is sound and practical weather training and decision-making that can prevent accidents. For instance, Knecht (2008a) found that pilots spent an average of 17-20 minutes reviewing information from weather products/providers prior to departure and 7 minutes reviewing information while enroute. However, the bottom 5% of the pilots sampled only spent 5-7 minutes on products/providers, and less than 2 minutes enroute. Whether or not relevant weather information can be extracted in such little time depends upon the nature of the weather and the end user's proficiency. However, it is somewhat disconcerting that something as critical as weather information could be allotted only a few minutes during pre-flight planning and even less in-flight.

Unfortunately, similar information was not obtained in this study. What we were able to determine was the type of weather information pilots possessed or had available during the flight. Specifically, archival AIRMET/SIGMET reports, weather radar data, and interview data made it possible to compare the information that pilots had versus the information actually available during the time of the encounter. An examination of the pilot responses in this study revealed that in most encounters there were more data sources available than the pilot actually accessed/obtained. In some situations, for example, the appropriate AIRMET/SIGMET was in effect along the route of flight and/or radar echoes were present, yet the pilot did not obtain them, or if they did, the information was misunderstood.

It would appear that knowledge and training about specific weather products and what information they can provide might prevent encounters with adverse weather from happening in the future. On the surface, this recommendation may seem obvious and perhaps redundant, given that weather-related training is required of all pilots before licensure. However, a review of training requirements for weather hazards reveals that private pilot applicants *"must receive and log ground training from an authorized instructor or complete a home-study course on … recognition of critical weather situations from the ground and in flight, windshear avoidance, and the procurement and use of aeronautical weather reports and forecasts"* (14 CFR 61.105). Even VFR-only pilots are given some instrument flight training as a precaution, so they can "maintain control of an aircraft while making a course reversal or diversion if they inadvertently enter clouds" (NTSB, 2005).

The issue is not whether or not a pilot should receive meteorological training – the regulations clearly state that they should and requirements are in place to ensure training is provided. The better question is "How does one adequately educate and train a pilot applicant to guarantee that the aforementioned requirements are met?" There appears to be no standard in the case of educating pilots on meteorology and training them to use weather analysis and forecast products properly. In fact, the only specific hazard discussed in the regulations cited above is wind shear. An interesting comparison can be made to stall recovery and avoidance training. During primary flight training, the student receives hands-on experience in stalls. This is straightforward, as stall training can be easily introduced. Practical introductions to managing weather-hazard areas cannot be as easily simulated.

One source of training materials and a useful resource is the Aeronautical Information Manual (AIM; FAA [2008]). The AIM includes a description of weather products as well as additional sources of weather information. However, while the AIM details many of these products, there are no specifics on what is required to be taught. It is unclear if pilots have been trained to interpret weather products adequately and make sound decisions based on their interpretation, or if they were taught just enough about each product to pass an examination. Should the latter scenario be the case, then it is possible that once an examination is over, a pilot may rarely consult some or all of these products in the future due to an incomplete understanding during initial training.

It is also interesting to note that a pilot can get all aviation weather questions wrong on the airman knowledge test (suggesting a total lack of weather knowledge), yet still pass the exam enroute to a private pilot's license (Wiegmann et al., 2008; NTSB, 2005). Likewise, during the required biennial flight review (BFR), "… the instructor giving the flight review is free to determine the content; therefore, the BFR may or may not include a demonstration of the weather knowledge and instrument flight skills required for initial certification" (NTSB, 2005).

Therefore, it is conceivable that after becoming certified a pilot would not be capable of interpreting risks associated with adverse weather and throughout their career may never be required to demonstrate knowledge on aviation-specific weather information products. Perhaps what is needed is either a refinement of the current airman knowledge test requirements, required weather-related items on the BFR, or some new approach to weather instruction like scenario-based weather training/testing. Maybe then, an improved understanding of weather and its implications while flying can be fostered throughout the GA community.

Finally, it would be disingenuous to suggest that all 25 adverse weather encounters were solely human factors-related, since more than half of the weather encounters were associated with some form of equipment failure. A sampling of the mechanical failures encountered by the pilots in this study included bad cells of an aircraft's battery; malfunctioning communication and navigation equipment like the radios, VOR (VHF Omnidirectional Range), GPS, and DME (Distance Measuring Equipment)). In each case, the aircraft was still operational; however, at a minimum, the equipment failure(s) produced a distraction but, in some cases, exacerbated the encounter with adverse weather.

CONCLUSIONS AND LESSONS LEARNED

Although the data presented here embodied the collective experiences of 25 GA pilots who encountered adverse weather in flight, the data have limitations. For instance, the sample was small and does not directly reflect the regional distribution of pilots within the U.S. Specifically, none of the pilots in this study were from the Western-Pacific or Central Regions. Furthermore, even though the largest number of participants came from the Northwest Mountain Region, none of the pilots were from Alaska where GA aviation is a prominent form of transportation. Likewise, roughly $3/4^{ths}$ of the pilots interviewed in this study held an instrument rating. This is similar to the Nall report's findings that slightly over half of the pilots involved in weather-related accidents were instrument-rated (AOPA, 2008).

It can also be argued that the pilots who volunteered to participate in our study did so because they felt that they simply made an honest mistake and did not willfully violate existing Federal Air Regulations (FARs). Whereas, the others who did not volunteer, may have declined because they knew that they were in violation and may have been reluctant to participate in an FAA-sponsored inquiry into why pilots fly into adverse weather. While it is impossible to know precisely why many pilots declined to participate (i.e., we did not ask them), it is possible that their participation would have increased the number of motivation-related encounters with adverse weather that may have driven the number of willful violations of FARs closer to those seen in previous GA accident studies.

Nevertheless, the data presented here do shed additional light on the conundrum of why a GA pilot would fly into adverse weather. Contrary to what the accident record seems to suggest, flight into adverse weather seems to be primarily due to the lack of appreciation/understanding of the hazards associated with adverse weather. Perhaps some encounters with adverse weather were motivated by outside influences or exacerbated by some manner of mechanical failure that may have led to the willful acceptance of unnecessary hazards. Simply put, there may be more to the story.

Furthermore, it would appear that additional effort should be placed in training, both *ab initio* and recurrent. Emphasis should be placed on ensuring a full understanding of the adverse impact of weather, including the recognition of IMC, icing, convective events, etc. Moreover, we recommend a standard weather briefing that includes multiple, converging sources of information.

In addition to analyzing the interview data to determine trends and patterns, it was also our intent to develop disciplined methods for analyzing future accidents and incidents. Ideally, this would enable investigators to better learn from these incidents, improve correction of involved personnel, and generate interventions. Specifically, we planned to generate a list of items and/or data points for investigators to gain a better understanding of what happened within a particular weather event. The structured interview and weather information gathered met that goal.

We had also intended to derive a simple standard protocol aimed at identifying relevant weather-related events that could be used by Flight Standards or other research teams, but this did not happen. Unfortunately, identifying cases by reviewing lists of flight assists and weather events listed in The Administrator's Daily Alert Bulletins was extremely time-consuming.

Selecting a set of potential adverse weather encounters was only the first step. A great deal of work remained to identify the appropriate FSDO, gain access to pilot contact information, contact the potential pilot-participant, and garner informed consent. However, in some cases, the FSDO did not have contact information or they did not have a record of the event. This was surprising because, according to the FAA's Air Traffic Quality Assurance process, flights assists are to be reported to the appropriate FSDO via the Regional Operations Center (ROC) within 3 hours of occurrence (7210.56C; FAA, 2002).

If the process of this study were to be duplicated or initiated as a proactive collection of near-miss data, the procedure would need to become formalized and streamlined across air traffic facilities and FSDOs. In addition, it would be useful if the standard flight assist form could be modified to archive key weather-related variables, many of which are identified within the structured interview used. Alternatively, the survey could be modified and streamlined for use by ASIs when conducting their investigations.

REFERENCES

Aircraft Owners and Pilots Association (AOPA) Air Safety Foundation (2004). Spatial disorientation: confusion that kills. Safety Advisor Physiology No. 1., Frederick, MD.

Aircraft Owners and Pilots Association Air Safety Foundation (2008) 2008. Nall report: Accident trends and factors for 2007. Retrieved from http://www.aopa.org/asf/publications/08nall.pdf

Beringer, D., & Schvaneveldt, R. (2002). Priorities of weather information in various phases of flight. *Proceedings of the Human Factors and Ergonomics Society 46th Annual Meeting.*

Beringer, D., & Harris, H. (2007). *Automation in general aviation: Two studies of pilot responses to autopilot malfunctions.* (Technical Report No. DOT/FAA/AM-07/24). Washington, DC: Federal Aviation Administration, Office of Aerospace Medicine.

Burian, B., Orasanu, J., & Hitt, J. (2000). Plan continuation errors: A factor in aviation accidents? *Proceedings of the International Ergonomics Association 14th Triennial Congress and Human Factors and Ergonomics Society 44th Annual Meeting, San Diego, CA.*

Code of Federal Regulations – Title 14: Aeronautics and Space, Part 61.65 – Instrument Rating Requirements. 14 CFR 61.65. Washington, DC: Federal Aviation Administration, December 2005.

Code of Federal Regulations – Title 14: Aeronautics and Space, Part 61.105 – Aeronautical Knowledge. 14 CFR 61.105. Washington, DC: Federal Aviation Administration, December 2005.

Detwiler, C., Boquet, A., Holcomb, K., Hackworth, C., Wiegmann, D., & Shappell, S. (2006). *Beyond the tip of the iceberg: A human factors analysis of general aviation accidents in Alaska and the rest of the United States.* (Technical Report No. DOT/FAA/AM-06/07). Washington, DC: Federal Aviation Administration, Office of Aerospace Medicine.

Federal Aviation Administration (2002). FAA Order 7210.56C Air Traffic Quality Assurance. Washington, DC: Federal Aviation Administration.

Federal Aviation Administration (2006). FAA Order 8000.1 Safety Management System Doctrine. Washington, DC: Federal Aviation Administration.

Federal Aviation Administration (2008). Aeronautical Information Manual – Official Guide to Basic Flight Information and ATC Procedures. Washington, DC: Federal Aviation Administration. February 14, 2008 (available on-line at http://www.faa.gov/airports_airtraffic/air_traffic/publications/atpubs/aim

FAR/AIM (2009). Federal Aviation Regulations/Aeronautical Information Manual. Aviation Supplies & Academics, Inc, 2008.

Goh, J., & Wiegmann, D. (2001). Visual flight rules flight (VFR) into adverse weather: An empirical investigation of the possible causes. *The International Journal of Aviation Psychology, 11*(4), 259-379.

Goh, J., & Wiegmann, D. (2002). Human factors analysis of accidents involving visual flight rules flight into adverse weather. *Aviation, Space, and Environmental Medicine, 73*, 817-822.

Knecht, W. (2005). *Pilot willingness to take off into marginal weather, Part II: Antecedent overfitting with forward stepwise logistic regression.* (Technical Report No. DOT/FAA/AM-05/15). Washington, DC: Federal Aviation Administration, Office of Aerospace Medicine.

Knecht, W. (2008a). *Use of weather information by general aviation pilots, Part I, quantitative: Reported use and value of providers and products.* (Technical Report No. DOT/FA/AM-08/06). Washington, DC: Federal Aviation Administration, Office of Aerospace Medicine.

Knecht, W. (2008b). *Use of weather information by general aviation pilots, Part II, Qualitative: Exploring factors involved in weather-related decision making.* (Technical Report No. DOT/FAA/AM-08/07). Washington, DC: Federal Aviation Administration, Office of Aerospace Medicine.

Macrae, C. (2009). Making risks visible: Identifying and interpreting threats to airline flight safety. *Journal of Occupational and Organizational Psychology, 82*, 273-293.

National Aeronautics and Space Administration Aviation Safety Reporting System (2007). General aviation weather encounters 2007. (Publication No. 63). Retrieved July 23, 2008, from http://asrs.arc.nasa.gov/docs/rs/63_ASRS_GA_WeatherEncounters.pdf

National Transportation Safety Board (2005). *Risk factors associated with weather-related general aviation accidents.* (Report No. SS--05-01). Washington, DC.

National Transportation Safety Board (2006, October). *NTSB safety alert: Thunderstorm encounters* (Publication No. SA-011). Washington, DC: Author. Retrieved from http://ntsb.gov/alerts/SA_011.pdf

O'Hare, D., & Owen, D. (2002). Cross-country VFR crashes: Pilot and contextual factors. *Aviation, Space, and Environmental Medicine, 73,* 363–366.

Patterson, J., & Shappell, S. (in review). Application of the human factors analysis and classification system to mining incidents/accidents in Queensland, Australia. Submitted to *Accident, Analysis, and Prevention.*

Reason, J. (1990). *Human Error.* New York: Cambridge University Press.

Shappell, S., & Wiegmann, D. (2009). Developing a methodology for assessing safety programs targeting human error in aviation. *The International Journal of Aviation Psychology, 19,* 252-269.

Stokes, A., Belger, A., & Zhang, K. (1990). *Investigation of factors comprising a model of pilot decision-making: Part II. Anxiety and cognitive strategies in expert and novice aviators* (Technical Report No. ARL-90-8/SCEEE-90-2). Savoy, IL: University of Illinois, Aviation Research Laboratory.

Wiegmann, D., Goh, J., & O'Hare, D. (2002). The role of situation assessment and the flight experience in pilots' decisions to continue visual flight rules flight into adverse weather. *Human Factors, 44,* 189-197.

Wiegmann, D., Boquet, A., Detwiler, C., Holcomb, K., Faaborg, T., & Shappell, S. (2005). *Human error and general aviation accidents: A comprehensive, fine-grained analysis using HFACS.* (Technical Report No. DOT/FAA/AM-05/24). Washington, DC: Federal Aviation Administration, Office of Aerospace Medicine.

Wiegmann, D., & Goh, J. (2003). Factors affecting pilots' decisions to continue flight into adverse weather. Paper presented at the *Annual Midyear Symposium of the American Psychological Association, Applied Experimental and Engineering Psychology Division,* Fort Belvoir, VA.

Wiegmann, D., Talleur, D., & Johnson, C. (2008). Evaluating weather-related training and testing of general aviation pilots. *Proceedings of the 52nd Annual Meeting of the Human Factors and Ergonomics Society,* New York, NY.

Wiegmann, D., & Shappell, S. (2003). *A human error approach to aviation accident analysis: The human factors analysis and classification system.* Aldershot, Great Britain: Ashgate.

Wiggins, M., Stevens, C., Howard, A., Henley, I., & O'Hare, D. (2002). Expert, intermediate and novice performance during simulated pre-flight decision-making. *Australian Journal of Psychology, 54,* 162–167.

Wiggins, M. & O'Hare, D. (2003). Expert and novice pilot perceptions of static in-flight images of weather. *International Journal of Aviation Psychology, 13,* 173-187.

APPENDIX A

A. Aircraft Information
   1. Aircraft Type
   2. Aircraft ownership (Own (full or part), rent, other)?
B. Basic Aircrew Information (Confirming demographic data)
   1. Age at time of incident
   2. Education (HS, College, graduate)
   3. Profession/Occupation
   4. Gender
   5. When did you first learn to fly?
   6. Total Flight Hours
   7. Flight Hours in Make/Model
   8. Flight Hours in the last 90 days
   9. Total Cross-Country Flight Hours (cross country defined as greater than 25nm from airport)
   10. Cross Country Flight Hours in last 90 days
   11. Ratings/Certificates
   12. Are you instrument rated? If so, how much instrument experience have you had (simulated and actual)?
   13. Do you participate in the WINGS program?
   14. Date of most recent flight instruction
   15. Year of most recent biennial flight review (BFR)
   16. Date of most recent medical exam
   17. Class of medical certificate (Class I, II, or III)
       a. Any waivers/limitations on the medical certificate
   18. If you used any of the following equipment, have you had training on it? If yes, what kind and how much?
       a. Communication Equipment
       b. De-icing Equipment
       c. Weather-avoidance Equipment
       d. Terrain-avoidance Equipment
       e. Autopilot
       f. GPS
       g. Additional Avionics (i.e., Altitude alert, radar altimeters)
       h. Multifunctional Displays (i.e. GPWS, TCAS)
C. Event Information
   1. Please describe the day of the flight.
   2. What was the purpose of the flight? (air tour, business, emergency medical, ferry, freight, passenger, pleasure, training)
   3. Have you made this trip before? If so, how many times?
   4. Preflight
       a. What is your normal method of preflight planning
       b. Was your method of preflight planning on the day of the incident different from normal?
       c. Did you attempt to obtain pre-flight weather information?
           1. If yes, when was the last time you checked weather before flight
       d. Did you have an alternate plan with alternate airports worked out in advance?
       e. Did you consider the geography along the flight route (e.g. mountains, rising terrain, etc.)?
       f. Which weather provider(s) were used?
           1. National Oceanic & Atmospheric Administration (NWS)
           2. Flight Service Station
           3. Direct User Access Terminal Service (DUATS)

4. Commercial Vendors
5. Hazardous In-flight Weather Advisory Service
6. Transcribed Weather Broadcast (TWB)
7. Pilot Automatic Telephone Weather Answering Service
8. En-route Flight Advisory Service
9. The Weather Channel
10. Other Pilots
11. Other (please specify):
    a. What type of information did you gather?

g. What weather providers(s) do you typically use?
   1. National Oceanic & Atmospheric Administration (NWS)
   2. Flight Service Station
   3. Direct User Access Terminal Service (DUATS)
   4. Commercial Vendors
   5. Hazardous In-flight Weather Advisory Service
   6. Transcribed Weather Broadcast (TWB)
   7. Pilot Automatic Telephone Weather Answering Service
   8. En-route Flight Advisory Service
   9. The Weather Channel
   10. Other Pilots
   11. Other (please specify):

h. Were any of your attempts to obtain pre-flight weather information unsuccessful? If yes, why?
   1. Did not know or were unable to find telephone or access numbers
   2. No telephone available
   3. No answer on telephone
   4. Telephone briefer did not have all requested information available
   5. Telephone briefer denied service
   6. No online access available
   7. Could not connect online
   8. Could not maintain online connection
   9. Required information not available on computer
   10. Experienced difficulty with computer interface
   11. Other (please specify)

i. What was the preflight weather forecast for the following?
   1. Departure airport (VMC/Marginal VMC/IMC)
   2. En-route (VMC/Marginal VMC/IMC)
   3. Destination airport (VMC/Marginal VMC/IMC)
   4. Did not obtain a preflight weather briefing
      a. If you did NOT attempt to obtain pre-flight weather information prior to departure, why not?
         i. Did not believe pre-departure weather was necessary
         ii. Was intimidated by process of obtaining weather
         iii. Did not know or was unable to find telephone or access number
         iv. No Telephone available
         v. No online access available
         vi. Other (please specify)

j. What was the planned length of flight?
k. Where did you depart from and when?
l. Where did you arrive and when?
m. What type of approach was planned for your destination? (CAT I, CAT II, CAT III)

n. Did you get a good night's sleep the night before?
o. How were you feeling on the day of the flight (e.g. sick, allergies, etc)?
5. In-flight
   a. What were the environmental conditions?
      1. Lighting conditions? (Day, Night, Dawn, Dusk, Cloudy, Sunny, Bright, Dark, etc…)
      2. Weather
      3. Temperature (degrees Fahrenheit)
   b. What was the weather like?
      1. At departure
      2. At destination
      3. During the Route
   c. Were you under any pressure to get to your destination?
   d. Did you have to deviate from the planned route or altitude?
      1. If yes, was it before or after you encountered weather?
   e. Were you aware of your course and location throughout the flight?
   f. Did you communicate with ATC?
   g. Did you communicate with any other en-route services (e.g. flight service stations)?
   h. What in-flight services did you request?
      1. Emergency climb/descent
      2. IFR clearance
      3. Vectors to an airport
      4. Vectors to VMC
      5. PIREPS
      6. Weather Updates
      7. IAP
      8. Other (please specify)
   i. Were there any distractions in the cockpit? (e.g., passengers, malfunctions.)?
   j. Was all the equipment working during the flight?
   k. What was your level of awareness of the weather (e.g., Did you know where the bad weather was headed?)
6. Upon encountering the weather
   a. During what point of the flight did you encounter weather?
      1. Was the pilot past the midpoint of the flight before realizing there was trouble?
   b. Was the actual weather better than, the same as, or worse than forecasted
      1. Departure (Better/Same/Worse)
      2. En-route (Better/Same/Worse)
      3. Destination (Better/Same/Worse)
   c. Can you explain your thought processes upon encountering the weather?
   d. Can you explain how you felt upon encountering the weather?
   e. Did you enter the weather?
      1. If yes, did you consider flying a 180?
   f. Have you ever experienced similar weather conditions in-flight?
      1. If yes, what did you do?

D. CONCLUDING THOUGHTS
   1. Is there anything that you feel we should know?
   2. What did you do that prevented this incident from being worse?
   3. Knowing what you know now, would you have done anything differently?

www.ingramcontent.com/pod-product-compliance
Lightning Source LLC
Chambersburg PA
CBHW081821170526
45167CB00008B/3496